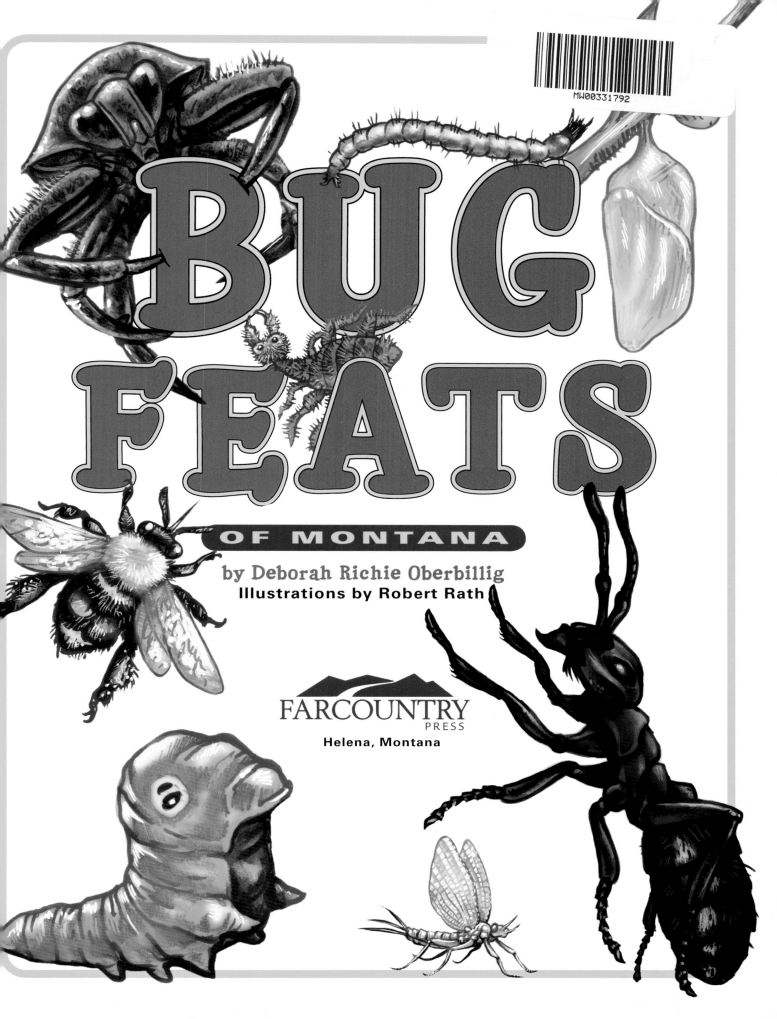

BUG FEATS
OF MONTANA

by Deborah Richie Oberbillig
Illustrations by Robert Rath

FARCOUNTRY
PRESS
Helena, Montana

Dedicated to August, Sam, Ian, Rebecca, Lucas, and Savanna

Acknowledgments
A huge thanks to the two best bug experts I could have hoped for to review this book:
Byron Weber: Elementary school teacher, bugster, naturalist, and Montana Public Radio bug commentator for "The Pea Green Boat."
John Acorn: Canadian bugster since age five, entomologist, president of the Lepidopterists' Society, author of entertaining bug books, and host of the international television series *The Nature Nut.*

Special thanks to Missoula's Rattlesnake Elementary School, teacher Catherine Schuck, and her 2007-08 fourth- and fifth-grade class of bug advisors: Kianna Aumiller, Gavin Booi, Ellie Brown, Evan Carol, Martin Chaney, Monica Fisher-Somerlott, Cree Folsom, Christopher Graef, Anastasia Halfpap, Peyton Humphries-McGovern, Courtney Kane, Bo Kendall, Brody Kendall, Carli Knox, Joey Manchester, Jackson McElroy, Anna McEvoy, Ian Oberbillig, Jake Oetinger, Madi Ruguleiski, Stephenie Ruguleiski, Mayah Van De Wetering, McKenzie Wingard.

Also thanks to Wease Bollman, Deb Fassnacht, Ellen Knight, Jen Marangelo, David Schmetterling, and the Watershed Education Network, as well as to editor Jessica Solberg and designer Shirley Machonis.

Photo credits
©2009 JupiterImages Corporation: pages 5 (compound eyes), 9 (spittle home), 22 (stonefly husks), 44 (ladybugs).
Eddie McGriff, University of Georgia, www.forestryimages.org: page 45 (adult green lacewing).
Whitney Cranshaw, Colorado State University, www.forestryimages.org: page 46 (forest).

ISBN 10: 1-56037-444-6
ISBN 13: 978-1-56037-444-2

© 2009 by Farcountry Press
Text © 2009 by Deborah Richie Oberbillig
Illustrations © Farcountry Press

For more information on our books, write Farcountry Press, P.O. Box 5630, Helena, MT 59604; call (800) 821-3874; or visit www.farcountrypress.com.

Library of Congress Cataloging-in-Publication Data

Oberbillig, Deborah Richie, 1958-
  Bug feats of Montana / by Deborah Richie Oberbillig ; illustrations by Robert Rath.
    p. cm.
  ISBN 978-1-56037-444-2
  1. Insects--Montana--Juvenile literature. I. Rath, Robert. II. Title.
  QL475.M9O24 2009
  595.709786--dc22
                        2008052490

Created, produced, and designed in the United States.
Printed in China.

14 13 12 11 10 09   1 2 3 4 5 6

TABLE OF CONTENTS

What's a Feat?

What do you call it when you jump high, run fast, climb a tree, or read a hard book? These are all "feats." This book is about bug feats—the fastest flyers, the loudest buzzers, and the sneakiest ambushers in Montana. Some of the bugs in this book are even world record holders!

dragonfly

What's a Bug?

This book uses the word "bug" to refer to insects, spiders, mites, scorpions, millipedes, centipedes, and their relatives. They all have "exoskeletons," hard coverings on the outside of their bodies that give them support and protection.

Bugs All Around

Bugs live in soil, air, water, snow, trees, caves, deserts, houses, and even on other animals. More than a million kinds of bugs are creeping, crawling, hopping, flying, and swimming on our planet. You can find the forty Montana bugs featured in this book in your backyard and schoolyard, or when camping, hiking, or floating a river.

What Do Bugs Eat?

Bugs eat meat, plants, or both. Some bugs even eat decaying plants and animals. Check out the Bug Finding Tips for each bug to find out if it is a predator (hunts for its meat), a plant eater, an omnivore (eats both), or a scavenger (eats decaying things).

Why Bugs Matter

Do you like chocolate? Or honey? How about strawberries? Many of the foods we eat depend on bugs. The plants need bugs to pollinate their flowers so they will bear fruit.

Bugs are food for fish, birds, and many other kinds of animals. Even grizzly bears eat bugs!

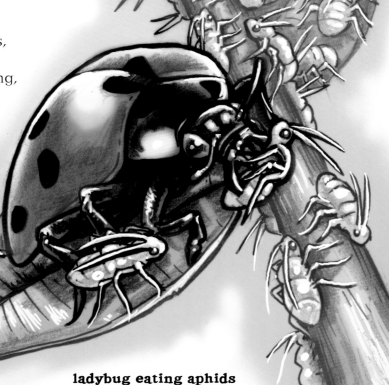

ladybug eating aphids

Bug Parts

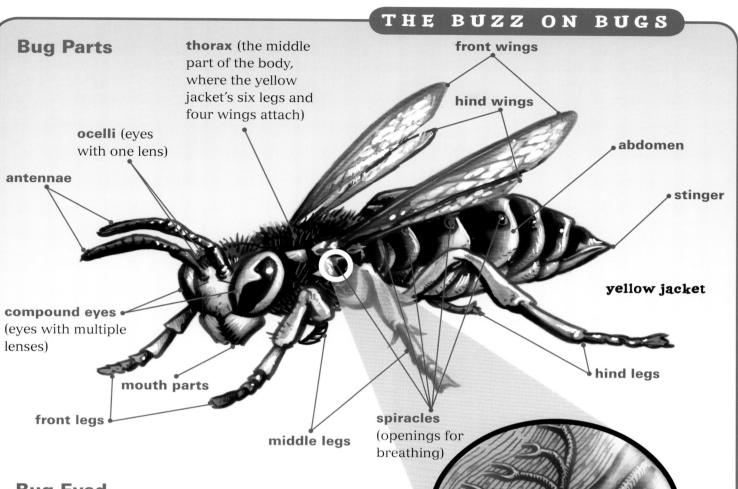

thorax (the middle part of the body, where the yellow jacket's six legs and four wings attach)

front wings

hind wings

ocelli (eyes with one lens)

abdomen

antennae

stinger

compound eyes (eyes with multiple lenses)

yellow jacket

mouth parts

front legs

middle legs

spiracles (openings for breathing)

hind legs

Spiracles look like this on the inside.

Bug Eyed

A bulging bug eye is chock full of mini-eyes. We call this a "compound eye." Humans have what is called a "single eye." With a single lens, you can easily focus your eyes to see near and far. Bugs can't do this. They can only see up close. Their compound eyes, however, are great for detecting fast motion over a wide area and picking up colors that are invisible to you.

compound eyes

Bug Breath

Bugs breathe through tiny openings along the sides of their bodies. These openings, called spiracles, have valves that open and close to control the flow of air. The reason many bugs are such great leapers and flyers is that they can take in more oxygen to fuel their muscles than you can by breathing through your lungs.

TRY THIS: To view the world like a bug, take a handful of drinking straws and hold them tightly in one hand. Then look through the cluster of straws.

A Bug's Life

When bugs grow up and change, it's called metamorphosis. Take a look at the two kinds of bug metamorphosis:

Complete Metamorphosis

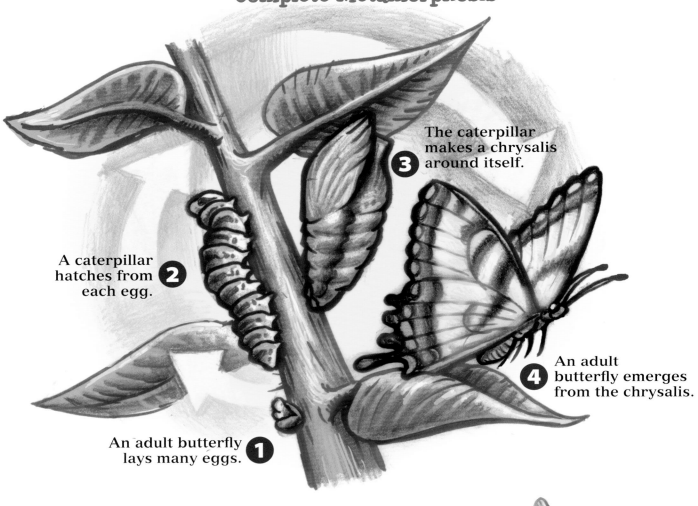

3 The caterpillar makes a chrysalis around itself.

2 A caterpillar hatches from each egg.

4 An adult butterfly emerges from the chrysalis.

1 An adult butterfly lays many eggs.

Incomplete Metamorphosis

  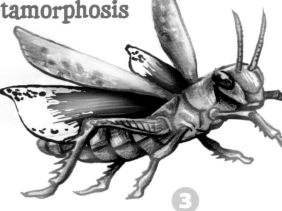

1 An adult grasshopper lays many eggs.

2 A wingless nymph hatches from each egg.

3 The nymph grows and molts and becomes an adult.

Cool Montana Bug Facts

How many? More than 20,000 different kinds of bugs live in the state.

Highest: Northern rock crawlers live on the tallest peaks, surviving freezing weather.

Dung-loving: A female Montana dung beetle tunnels underneath a pile of poop, pulls the dung into her tunnel and packs it down, lays her egg in it, then covers it up neatly, leaving a space around the egg for the larva to grow. What does the larva eat? Dung, of course.

dung beetle

Grizzly Bite: Army cutworm moths (also known as "millers") fly up to mountaintops in late summer to find nectar, but they also meet up with grizzly bears, which scoop them up by the thousands—yum.

army cutworm moth

BUGSTER TIPS

A bugster enjoys looking for bugs and spends lots of time outdoors on warm days, when bugs are most active. Here are a few bugster tricks of the trade:

Walk slowly, stop often, and crouch down.

Listen for buzzes, trills, and hums.

Look for clues such as chewed-up leaves, holes in wood, spiderwebs, and anthills.

Get up close to flowers, bushes, trees, and the water's edge.

Attract insects by turning on a light at night.

bug journal

butterfly net

hand lens

collecting jar

LONG JUMP CHAMP
Grasshopper

A grasshopper can jump twenty times its body length. If you could magnify a grasshopper to your size, you'd want this insect on your track and field team. But there's one problem. A hopper doesn't always land right side up and hops in any direction.

When out fishing one day, a robot designer noticed that grasshoppers fell on their sides, righted themselves, and leaped again. Inspired, he created a new robot for exploring the surface of Mars. Where once a wheeled robot got stuck every time a rock blocked its way, the improved robot now hops, crashes, turns right side up, and jumps again.

Snack Fact

More than 400 hopper species live in the seventeen western states. Many birds snack on grasshoppers, including a small falcon called a kestrel. Fox kits and coyote pups learn to hunt by pouncing on the tasty hoppers.

How Far? Try This

Make a chalk mark in the middle of the floor or outside in your driveway. Place a grasshopper (a young one, without wings) on the mark. Watch it jump and mark the landing. Then measure how far the hopper leaped. Divide that distance by the body length. How far can you jump compared to your height?

(Hint: If a grasshopper is 1 inch long and jumped 20 inches, that's 20 divided by 1 = 20 body lengths.)

BUG-FINDING TIPS

Key features	✓ has big hind legs for jumping ✓ nymphs look like adults but are wingless ✓ has many patterns ✓ can be the size of your little finger ✓ plant eater
Where	In grass.
When	Spring through fall. Can find nymphs and adults together from midsummer on.
Listen for	Chirps, trills, and buzzes; wings clack in flight.

What's a grasshopper's favorite year?

(Answer: A leap year!)

HIGH JUMP CHAMP
Spittlebug

A spittlebug adult is also called a "froghopper," a perfect name for this leaper that can jump two feet in the air—100 times higher than its height. If a spittlebug were the size of a pro basketball player, it could jump 500 feet above the basket! If you had this bug's superpowers, you could bound over the famous arch in St. Louis or to the top of a seventy-story building. The spittlebug has the high jump record for its size, beating out another terrific hopper—the flea.

Spittle Home

The nymph of the high-leaping adult lives inside a bubbly home on a plant stem. If you could be this nymph for a day, you'd need spittle-making instructions:

Stand on your head. Insert your mouth into a plant and suck up plant juice until you have extra liquid that will shoot out the rear of your abdomen. Raise and lower your abdomen tip to open and close plates on your underside, frothing up the juice. Keep it up until you're covered in wet foam. This takes about twenty minutes.

Home Sweet Home

spittle home

Spring-Loaded Legs

A spittlebug folds its short legs under its thorax and locks them into place. When a predator pounces, the spittlebug escapes by unlocking its legs and springing up into the air.

BUG-FINDING TIPS

Key features	✓ a squat leaper ✓ has green or brown patterns ✓ the length of your little fingernail ✓ plant eater
Where	Weeds, tall grass, shrubs, or pine twigs (if a pine spittlebug).
When	Summer.
Look for	**Adult:** Crouch down in the grass and wait for one to hop on you. **Nymph:** A bubbly spittle home on a plant stem.

FASTEST LAND BUG
Tiger Beetle

On your mark, get set, go! The tiger beetle races across the sand to the finish line. It traveled twenty-nine body lengths per second. A grizzly bear would have to run at 135 miles per hour to match the effort.

Tiger beetles depend on swiftness to hunt down their insect prey. When they reach top speeds, they lose their vision, so they have to stop, regain their sight, then sprint forward again.

"With bulging compound eyes, she scans the open ground for prey. Then, at incredible speed, she rushes forward. Whoa! Hard to see when you are running that fast! She slams on the brakes for a fraction of a second, gets her bearings and redirects the charge. This time, the intended victim turns out to be nothing but a dark plant seed rolling down the slip face of the dune. It grates against her long, toothy jaws. The next target, however, is a wandering cutworm, and her mandibles sink deep."

—JOHN ACORN,
TIGER BEETLES OF ALBERTA

CICINDELOPHILY!

Tiger beetle followers have a case of "cicindelophily" (sis-in-del-OFF-illy), the love and admiration of tiger beetles.

Hit the Trail

At least twenty-three kinds of tiger beetles live in Montana. To find them, look for nature's racetracks: beaches, sandbars, dunes, well-worn trails, and mudflats. Beetles need a flat surface where they can run without obstacles and see what is ahead of them. Unfortunately, their habitat also can attract people with dirt bikes and off-road vehicles, which are deadly to the beetles. Tiger beetles show us why it's important to value all places in nature.

BUG-FINDING TIPS

Key features	✓ big eyes and jaws ✓ long antennae ✓ long legs ✓ colorful markings ✓ half an inch long ✓ predator
Where	Beaches, sandbars, dunes, mudflats, and well-worn trails.
When	Summer.
Look for	On bare trails, look for beetles that run in front of you, fly up, then land a few feet away.

MASTER AVIATOR
Western Meadowhawk Dragonfly

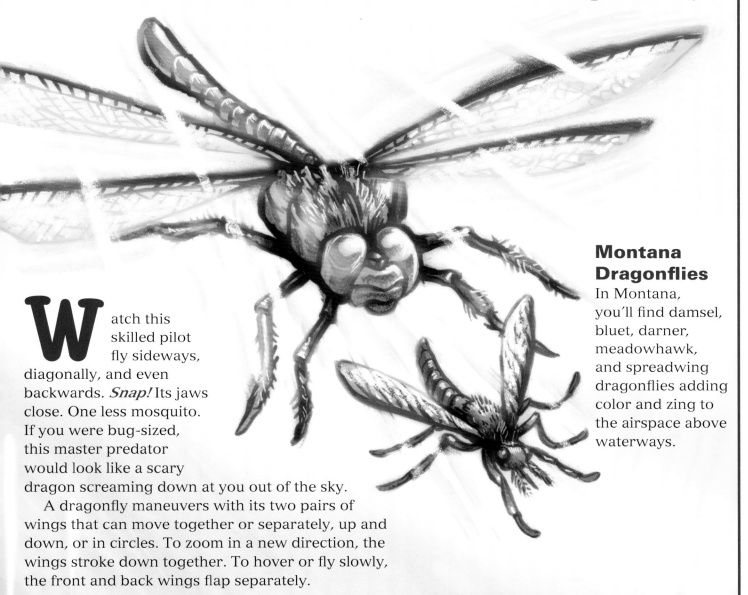

Watch this skilled pilot fly sideways, diagonally, and even backwards. *Snap!* Its jaws close. One less mosquito. If you were bug-sized, this master predator would look like a scary dragon screaming down at you out of the sky.

A dragonfly maneuvers with its two pairs of wings that can move together or separately, up and down, or in circles. To zoom in a new direction, the wings stroke down together. To hover or fly slowly, the front and back wings flap separately.

Montana Dragonflies

In Montana, you'll find damsel, bluet, darner, meadowhawk, and spreadwing dragonflies adding color and zing to the airspace above waterways.

Fastest Insect?

Some believe dragonflies hold the flight speed record, at thirty-six miles per hour. However, it's tough to reliably measure insect speeds because so many factors affect how fast a bug flies, such as temperature, humidity, wind, or whether it is flying freely or in a laboratory. So we're still waiting for someone (like you?) to measure the truly fastest flying insect.

BUG-FINDING TIPS	
Key features	✓ orange-brown on lower half of each clear wing ✓ about an inch and a half long ✓ predator
Where	Ponds, lakes, and marshes.
When	Summer to fall. Most active on sunny, calm days.
Look for	Dragonflies perching on rocks or branches.

WHIRRING WINGS
Mosquito

Zzzzzzzzzzz. That high-pitched whine comes from a mosquito beating its wings as fast as 500 times per second. High-speed, noisy wings have another purpose beyond flying. A male mosquito "listens" for whirring females with his feathery antennae. He identifies a female of his own species by her wing sound. Each species of mosquito makes its own special sound by beating its wings faster or slower.

SKEETER FACTS

✓ Only female mosquitoes bite. They need blood to get the protein to produce eggs.

✓ To find their prey (you!), female mosquitoes detect the carbon dioxide in your breath. Hold your breath and you'll be fine—until you exhale!

✓ Male mosquitoes pollinate flowers.

mosquito larva

What insect has the fastest wing beat record?
A no-see-um (a tiny biting midge) whirs its wings at 1,046 times per second.

A Mosquito Meal
Slap, slap . . . scratch. Sometimes mosquitoes dine on you, and sometimes mosquitoes are a meal for other creatures. Diving beetles, dragonfly and salamander larvae, and small fish eat mosquito larvae wriggling in the water. Birds, bats, and dragonflies dine on the flying adults.

What is a mosquito's favorite sport?
(Answer: Skin diving!)

BUG-FINDING TIPS

Key features	✓ two wings
	✓ slender body
	✓ long legs
	✓ less than half an inch in length
	✓ females are predators
	✓ males are plant eaters
Where	Near water: puddles, lakes marshes, and wet meadows.
When	All summer, with peaks in midsummer and after rains. Most active at dusk and dawn.
Listen for	High, whining Zzzzzzzzzzz!

World RECORD HOLDER

LOUDEST BUZZER
Cicada

An orchestra timpanist strikes the timpani (two or more kettledrums) with sticks to produce deep sounds. The cicada (*sick-AY-duh*) has its own built-in kettledrums, called "tymbals," located on either side of its body. Instead of a stick, the cicada uses a large muscle to vibrate the tymbal to make a dry, buzzing *SSSSSSSSS* that's amplified in its air-filled abdomen (its drum).

Keep It Down!

If you stand a foot away from a cicada making its loudest alarm call, it can be louder than a jackhammer! We measure sound in decibels. A jack-hammer is 90 decibels. The loudest cicada in North America can buzz at 105.9 decibels. The African cicada holds the world record for the loudest insect noise at 106.7 decibels.

What do you call a musical bug?

(Answer: A humbug!)

DRIVEN TO DRUM
Males attract mates and rivals with their buzzing drums.

BUG-FINDING TIPS

ey features	
	✓ blackish in color
	✓ about one inch long
	✓ wings at rest are held rooflike over its body
	✓ wingless nymph lives underground for several years
	✓ plant eater (larvae only; adults do not feed)
Where	Habitats with trees.
When	Summer (July/August peak). Most active in the heat of the day.
Look and listen for	Rattling buzz that grows louder then softer; search thin branches of trees overhead.

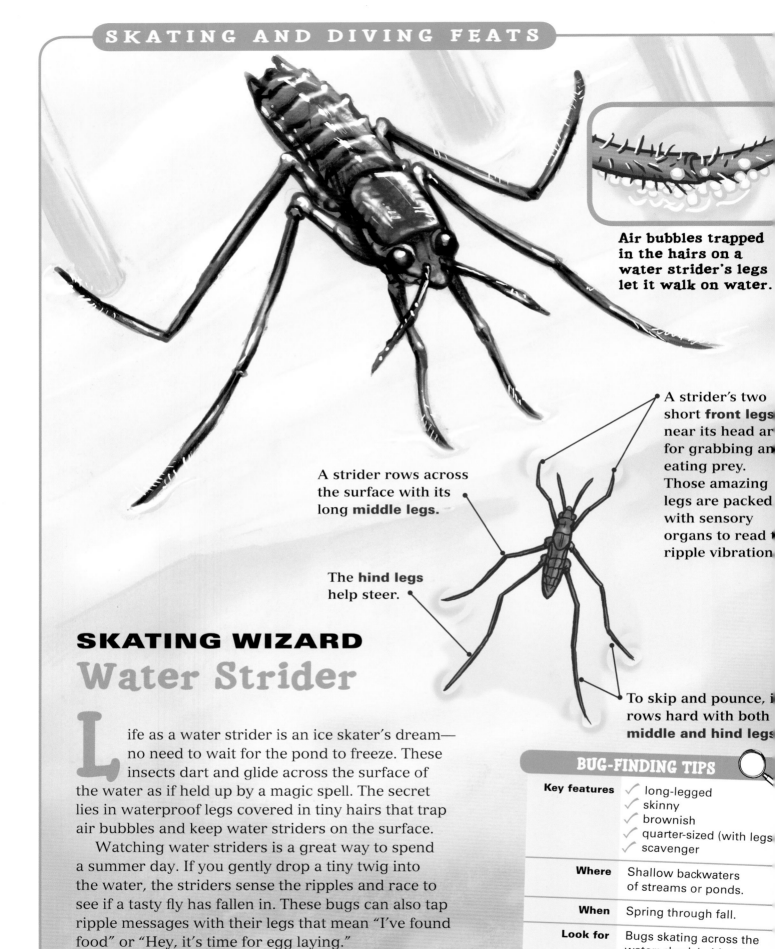

Air bubbles trapped in the hairs on a water strider's legs let it walk on water.

A strider's two short **front legs** near its head are for grabbing and eating prey. Those amazing legs are packed with sensory organs to read the ripple vibrations.

A strider rows across the surface with its long **middle legs.**

The **hind legs** help steer.

To skip and pounce, it rows hard with both **middle and hind legs.**

SKATING WIZARD
Water Strider

L ife as a water strider is an ice skater's dream— no need to wait for the pond to freeze. These insects dart and glide across the surface of the water as if held up by a magic spell. The secret lies in waterproof legs covered in tiny hairs that trap air bubbles and keep water striders on the surface.

Watching water striders is a great way to spend a summer day. If you gently drop a tiny twig into the water, the striders sense the ripples and race to see if a tasty fly has fallen in. These bugs can also tap ripple messages with their legs that mean "I've found food" or "Hey, it's time for egg laying."

BUG-FINDING TIPS	
Key features	✓ long-legged ✓ skinny ✓ brownish ✓ quarter-sized (with legs) ✓ scavenger
Where	Shallow backwaters of streams or ponds.
When	Spring through fall.
Look for	Bugs skating across the water; check behind fallen logs in streams.

SCUBA DIVER
Water Boatman

How does the water boatman stay under water for up to six hours? This ingenious bug traps an air bubble at the surface, tucks it within the water-repellent hairs on its belly, and then heads down like a scuba diver carrying a full tank of oxygen. Gradually, the air supply dwindles as the nitrogen in the bubble dissolves.

BOATMAN WATCHING

Look down at the edge of a pond or slow-moving stream to see a water boatman rise to the surface with the tip of its abdomen pointing up to catch a new air bubble. Watch a boatman hang onto vegetation beneath the water in order to keep from floating up to the surface when it's not ready yet (the buoyant air bubble tends to pull the insect to the surface).

front legs: for shoveling up tasty decaying bugs in the algae

middle legs: for hanging on tight to plants or rocks

spiracles (vents on side of abdomen and thorax): for taking in oxygen

wings: leathery wings for flying to find new places to live when home waters start drying up

air bubble: placed under its abdomen for a supply of air

back legs: for rowing

BUG-FINDING TIPS

Key features	✓ swims right side up ✓ oarlike hind legs ✓ half an inch long ✓ scavenger
Where	Ponds, still waters, and large puddles.
When	Summer, fall.
Look for	Fast, darting swimming or resting on underwater plants; flies to porch lights at night.

BEST FLIP
Click Beetle

A click beetle lying on its back looks helpless. You reach to help turn it over, but before you can— *click*—the beetle explodes into action. It makes a *click* sound as it flips over and lands right side up. Although flipping for a living sounds like fun, it's also a lifesaving device for this beetle to avoid hungry predators.

Built to Flip

If a click beetle gets turned over on its back, it uses a special trick to flip back over. A hinge between its thorax and abdomen allows the beetle to arch up, so only its ends are touching the ground. What it's doing is snapping its long spine into a groove, which loads the spring. When it straightens back out, the spring-loaded beetle flips with a loud *click.* Unfortunately, it only lands right side up about half the time.

BUG-FINDING TIPS

Key features	✓ slim and dark ✓ has grooves running down wing covers ✓ rounded head ✓ the size of a small paperclip ✓ some kinds have fake eyespots ✓ plant eater
Where	On plants, the ground, decaying wood, moss, and under bark.
When	Summer. Active during the day.
Look for	A click beetle by itself; they don't live in groups.

LOWLY BEGINNINGS

This wireworm larva that wriggles around underground will transform into the acrobat of the beetle world. Like butterflies, all beetles go through complete metamorphosis—from egg to larva to pupa to adult (see page 6).

click beetle larva

WEIGHT-LIFTING WONDER
Wood Ant

"An ant on the move does more than a dozing ox."
—Lao Tzu

A tiny ant can pick up a pebble fifty times its own weight. To see these weight lifters in action, visit a nest mound of the wood ant on a warm summer day. Watch ants hauling seeds, pine needles, or other mighty loads.

Although ants are strong, they are not the strongest insects. The rhinoceros beetle (up to six inches long and found in warmer climates than Montana) is the strongest creature on earth for its size—able to lift 850 times its own weight.

What is an ant's favorite song?
(Answer: The National Ant-them!)

Mound Maker

Ants do a lot of lifting to build a mound, hauling countless sticks, pebbles, or needles uphill to make a lofty mansion several feet tall. The round mound captures more sunlight than a flat shape, and the interior can be up to 20 degrees warmer than the outside. Inside, thousands of ants scurry through a maze of tunnels. Their complex society consists of sterile females (can't reproduce) that serve as workers or soldiers, winged fertile males (can reproduce) called "drones," and one or more queens.

TRY THIS: Take a felt tip marker and color a pine needle or small twig. Place it on the edge of an ant mound that you can visit every day. Watch to see where the ants move it.

BUG-FINDING TIPS

Key features	✓ a red head and thorax ✓ black abdomen ✓ fingernail length ✓ builds mounds ✓ omnivore
Where	Forests, meadows, and weedy backyards.
When	Spring, summer.
Look for	A flicker (a woodpecker) hunting for ants on a mound; lines of ants coming and going.

BEST CAMPER
Tent Caterpillar

Tent caterpillars spin their own silken tent that's solar-heated, waterproof, windproof, and expandable. They choose the best spot, tighten the rain fly, stay comfy in the tent, and know the way back to camp. Here's how they do it:

Choosing the Best Spot:
Morning sunshine feels great when you're camping. Tent caterpillars know which parts of a branch will warm up first after a cool night—that's where they put up the tent.

Tightening the Rain Fly:
The caterpillars pull each silken waterproof strand extra tight. A taut rain fly keeps the rain from pooling on the tent.

Never Lost:
Trailing a silken thread behind them, caterpillars head out to find green leaves to eat. They also lay down invisible scent trails (called "pheromones," *FAIR-uh-moans*) to share messages about the best feeding grounds. To survive, these caterpillars need each other, so it makes sense to share good feeding spots.

Heating and Cooling:
Instead of snuggling in sleeping bags, they build layers inside the tent and move around. When it's cold, they press against the outer wall closest to the sun. When it's warm, they crawl to the middle layers to get cool. When it's really hot, they hang outside the tent and dangle in the breeze.

BUG-FINDING TIPS

Key features	✓ reddish brown with a row of blue and orange spots ✓ about one inch long ✓ adult moth is stout and brown ✓ plant eater
Where	Forests with leafy trees and shrubs.
When	Spring, summer (some years more than others).
Look for	White shiny webs in branches. Birds devour the caterpillars and pluck silken strands from the tents to line their nests.

FIRST PAPER MAKER
Bald-Faced Hornet

The Chinese invented paper more than 2,000 years ago. But were they first? Ask a bald-faced hornet—a wasp that has made paper nests for millions of years. Here's what a queen hornet might tell you if she could:

"I start the nest when I wake in spring. First, I bite off a chunk of tree bark and chew it up with my starchy saliva to make pulp. I find a sturdy branch and spread the pulp with my legs and jaws into a carton-like shape. The pulp dries into paper. I fill the carton with six-sided paper cells, where I will lay my eggs. I measure each cell with my antennae so they will all be the same size. When my eggs hatch, I feed them until they grow into workers, who will finish the nest."

A football-sized paper nest can hold 200 hornets.

BUG-FINDING TIPS

Key features	✓ black and white ✓ one inch long ✓ omnivore
Where	Nests are found in shrubs (sometimes low to the ground) and trees and under house eaves.
When	Summer, fall. In winter, you can see empty nests in leafless trees.
Look for	Workers guarding the nest; a torn-up nest is a clue that a bear, bird, or raccoon ate the larvae in the cells.

TRY THIS:
If you see hornets making a nest, leave out a piece of old, flaky painted wood. Watch to see if they shred the wood for paper and swirl the colors into the nest. Keep your distance —one hornet can deliver several nasty stings to protect its nest.

The Four Cs: Clear, Cool, Clean, Connected

Montana is famous for trout fishing in rivers such as the Missouri, the Gallatin, and Yellowstone. Native trout eat aquatic bugs that in turn need healthy streams and rivers to survive. What makes a healthy trout stream or river? Just remember the *Four Cs: clear, cool, clean, and connected.*

Streams stay *clear* and *cool* when trees shade the water and roots keep banks from collapsing. *Clean* streams are free of pollutants. *Connected* streams link to bigger streams so that fish can swim freely from one to the other. This section features the feats of three bugs that are the main indicators of healthy streams and rivers: the caddisfly, mayfly, and stonefly—plus a fourth clean-water-loving bug, the damselfly.

BEST DECORATED HOUSE
Caddisfly

Pick up a rock in a cool, clear stream. Turn it over and you may find a caddisfly larva living in a house decorated with pebbles or twigs and moss. Look closely for a head, thorax, and legs popping out at one end. Always place the rock back in the water the way you found it. A caddisfly larva weaves a tube of saliva and silk, then glues on pieces of plants, twigs, sand, pebbles, or shells. Each kind of caddisfly selects its own special building materials.

caddisfly larva

BUG-FINDING TIPS

Key features	✓ tube cases about half an inch long ✓ plant shredders, grazers, or predators
Where	Under rocks in healthy streams.
When	Larvae seen year-round; adults hatch in spring.
Look for	Tube cases. Generally, you can tell one species from another by the house it builds, like twigs versus pebbles.

SHORTEST ADULT LIFE
Mayfly

Mayflies swarm above the river like smoke from a campfire. Trout gobble up the hatching insects. The mayflies that escape must quickly find mates, and the females must lay eggs before dying. Their partially formed mouths would not allow them to eat even if there was time for a meal. We call mayflies "ephemeral" *(ee-FEM-uh-ruhl)*, meaning short-lived. An adult mayfly in Montana lives fewer than two days.

Life Story

Mayfly nymphs breathe underwater through gills. Some species filter food that drifts by in the current. Others scrape plants from rocks. They may live one to two years before hatching into partial adults called "duns" that fly up to shore and shed their skins to emerge as true adult mayflies.

YELLOWSTONE NATIONAL PARK COOL FACT

More than sixty kinds of mayflies live in the Gallatin River alone. They have names like pale morning dun and gray drake.

What comes after a Mayfly?
(Answer: A junefly!)

NYMPH-FINDING TIP:

Most have three tails—remember it this way: hold three fingers upside down to form an M for mayfly. Also look for fuzzy gills on the sides of the abdomen.

BUG-FINDING TIPS

Key features	✓ most winged mayflies have three tails ✓ about three-eighths inch long ✓ plant eater (nymphs only; adults don't eat)
Where	Healthy streams and rivers.
When	Spring to late fall (large hatches in fall when leaves turn gold, some in spring).
Look for	Mayflies dancing in the air above the water.

BEST DRUMMER
Stonefly

What's a drummer without sticks or hands? A stonefly. This river insect drums with its belly, or abdomen. A male beats out a pattern to call to a female: *ba da da DA DA DA DUM.* She responds with a quieter *ta—ta—ta—ta—ta.* If all goes well, they find each other and mate.

Scientists recorded the drumming of 150 species of stoneflies and made a remarkable discovery. Every species has its own special pattern, which helps a male meet a female of its own kind. Unfortunately, the drumming is too soft for us to hear.

MONTANA STONEFLIES

✓ Montana has 113 species of stoneflies that fall into two groups: predators or plant shredders.

✓ The common perlid stonefly swallows small aquatic bugs whole.

✓ The giant stonefly (salmon fly) is the cow of the stream—a shredder.

✓ Glacier National Park's McDonald Creek is excellent stonefly habitat.

TRY THIS: Divide into pairs. Come up with a unique drumming pattern and drum back and forth like two stoneflies. Practice by tapping your hands on a table—fast, slow, loud, soft.

stonefly husks

BUG-FINDING TIPS

Key features	✓ can be pink, earning the nickname "salmonfly" ✓ can be two inches long (giant stonefly adult) ✓ plant eater
Where	Healthy streams, rivers with rocks, and streamside shrubs with branches for adults to climb on.
When	Spring is best for hatches. June is known for giant stonefly (salmonfly) hatches.
Look for	Shed skins of emerging stoneflies on rocks by a river. Turn over rocks in the water to find a nymph with two long tails.

FLYING TOOTHPICK OF DOOM
Bluet Damselfly

The mountain lake's clear waters glisten in the sunshine. At the shallow end, among the reeds, you notice a swarm of what look like electric-blue, flying toothpicks. To you, they're probably strange and beautiful. To bugs, these damselflies are terrifying.

Like all damselflies, they zoom after prey on speedy wings. Their large eyes are packed full of lenses and can pinpoint tiny insects several yards away. Bluets can fly slowly among grasses, too, dipping down to pluck up an aphid or other treat.

Oh no—the Claw . . .
Underwater, the nymph damselfly might be the called the "Swimming Claw of Doom." The nymph chases after mosquito larvae and other delights. When it gets close enough to strike, it shoots out a clawed lower lip and snags the prey.

Damselfly Drama
Damselfly watching is never dull. See if you notice behaviors like two damselflies hooked together in flight. What's that all about? A male is mating with a female and stays locked with her until she lays the eggs that he has fertilized. That way, he ensures that she will lay his eggs and not some other male's eggs.

BUG-FINDING TIPS

Key features	✓ males are bright blue with black spots ✓ females can be greenish yellow ✓ toothpick-sized ✓ predator
Where	Ponds, lakes, and slow-moving streams.
When	Summer (adults), year-round underwater (nymphs).
Look for	Bluets hovering among the reeds in shallow water.

SMOKE CHASER
Black Fire Beetle

Lightning strikes a dead fir tree. *Crack!* Flames race through the forest on a hot August day. Fifty miles away, a female black fire beetle detects the heat with her pair of heat sensors. Off she flies, searching out the smoke with her antennae. The beetle whirs around a smoldering tree, looking for a crevice in the bark to lay her eggs. Those eggs will hatch into juicy, white larvae that tunnel into the wood for a year before pupating and emerging as winged adults. But first they have to escape the barbed tongue of another fire-loving critter: the black-backed woodpecker.

Heat Sensors: A Closer Look

On either side of the beetle's thorax are pits that hold about seventy domes. These domes detect heat from a faraway fire that travels as invisible rays called "infrared." The infrared rays strike each dome and warm up a ball inside. This ball expands and bumps into a nerve cell that sends a clear message to the beetle: "fire and smoke!"

What do you call a beetle that rocks?
(Answer: John Lennon!)

heat sensor

What other Montana creature has heat sensors?

A rattlesnake has two pits on its head that detect heat so the snake can find mice and other warm-blooded animals to eat. But those heat sensors could never find a mouse miles away!

BUG-FINDING TIPS	
Key features	✓ a type of metallic woodborer ✓ the size and color of a shiny black bean ✓ plant eater
Where	Recently burned evergreen forests.
When	Summer.
Look for	Burned trees with holes or tunnels in the back from burrowing larvae.

COLOR CHANGER
Goldenrod Crab Spider

A bee buzzes toward a white daisy, ready to enjoy a nectar meal. But this flower holds a predator in disguise. Today, the goldenrod crab spider is white to match the daisy's petals. A week later, she will be yellow to hide in a nearby sunflower. The bee lands. The spider pounces. She enfolds her victim with her long front legs and plunges in her fangs. The venom paralyzes and then dissolves the bee's insides. The spider sucks up her liquid meal. Bee-licious?

Color Changer

A female starts out white. After a spider scuttles crablike up a stem and nestles into a yellow flower, it takes several days to change color. The reflected yellow from the flower sends a message to produce a yellow pigment. First the abdomen changes, then the cephalothorax and legs turn yellow. Only females can change from white to yellow and back to white.

YOU ARE WHAT YOU EAT

A baby crab spider that eats a red-eyed fruit fly turns pink for a few days, then changes back to white. Only a young spider can change color by eating, not an adult.

BUG-FINDING TIPS

Key features	✓ crablike ✓ holds its legs outstretched to the side ✓ scuttles sideways and backwards ✓ the size of your little fingernail ✓ predator
Where	Gardens, meadows, fields near woods.
When	Summer.
Look for	White or yellow flower blossoms like daisies or goldenrods.

ANTIFREEZE INVENTOR
Snow Flea

Imagine if your body could keep you from freezing on the coldest days. No more jackets! The tiny snow flea, also called a "springtail," does just that by making antifreeze to keep the liquids in its body from turning to ice.

The snow flea's antifreeze protein may hold the secret to saving hearts for transplants. Right now, it's a race against time to preserve a heart on its way to a waiting patient. The heart needs to be as cold as possible without freezing and then warm up when transplanted. That's why scientists study the snow flea's protein; its antifreeze keeps the liquids in its body from freezing, and then, when it's warm, that antifreeze breaks down and lets the liquids warm up again. This special quality makes the tiny snow flea a giant in the medical world.

OLDEST KNOWN BUG

A paleontologist found a 400-million-year-old springtail fossil in a rock in Scotland.

JUMP FOR JOY

The snow flea is part of a big group of bugs called springtails, named for their springy tails. To escape danger, this wingless bug catapults high into the air. The trick lies in a forked, tail-like structure that it latches tight to its belly. To spring, the prongs let go; the tail pushes against the ground and shoots the snow flea into the sky.

BUG-FINDING TIPS

Key features	✓ dark purple ✓ very small, the size of ground pepper ✓ scavenger
Where	On the snow or on leaves on the ground.
When	Winter (warmer days when temperature is just above freezing).
Look for	Black dots on the snow. Ski tracks and footprints in the snow may collect many snow fleas.

SNOW-DAY FLIER
Mourning Cloak Butterfly

Montana State Butterfly

"Dreams are brought to us in our sleep by a butterfly."
—*Blackfeet belief*

This butterfly hibernates in tree cavities or old buildings and wakes in March to flit about on the last of the warm yet snowy days of winter. Montana kids love this time of year and will race around in short-sleeved shirts and even shorts. No wonder that in 2001 students picked the mourning cloak to be the state butterfly.

Staying Warm

Do you feel warm when wearing a black shirt in the sun? Dark colors absorb heat, and light colors reflect heat. Look for the dark-colored mourning cloak butterfly on a tree trunk or rock with its wings opened and angled toward the sun. Like the snow flea, the mourning cloak also produces antifreeze—sugarlike chemicals that prevent damaging ice crystals from forming inside its cells and tissues.

WHAT'S IN A NAME?

People once wore long, dark cloaks when mourning for someone who died. This butterfly's coat may be somber, but a butterfly in March is a cheery sight after a long winter.

BUG-FINDING TIPS

Key features	✓ has dark maroon wings trimmed in yellow with blue spots ✓ three inches wide (the size of your palm) ✓ feeds on nectar and sap
Where	Every county in Montana; in forest openings, parks, and gardens, and along rivers and streams.
When	March to mid-November, except August, when this butterfly rests.
Look for	Tree with woodpecker or beetle holes or other wounds that drip sap and attract this butterfly.

BREAK-IN EXPERT
Box Elder Bug

Squeezing through tiny cracks, the box elder bug breaks into your house for a cozy winter hideaway. When it's too chilly outdoors for bugging, you can watch one of these uninvited guests crawling across your homework. Put a jar over one so you can take a closer look. Take one to school for show-and-tell. If you aren't allowed to let it go again in your house, find a deep crack in the bark of a tree for it to hide.

HOUSE GUEST OR PEST?

Adults tend to think any bug crawling around the house is a pest. You might point out to them the good qualities of this gentle houseguest:

✓ Not a stink bug—in fact, not smelly at all.

✓ Not hungry—will leave the house plants alone.

✓ Not a biter.

✓ Not full-time—flies away in spring.

Name Game

The box elder bug is called a "maple bug" in Canada. Both are named for the same kind of tree that this bug hangs out in during summer.

BUG-FINDING TIPS	
Key features	✓ black with red markings ✓ red abdomen ✓ up to half an inch long ✓ nymph is wingless ✓ plant eater
Where	Outdoors, feeding on seeds of box elder, maple, and ash trees.
When	Spring (on south side of buildings), summer (in trees), and fall/winter (may come indoors for shelter).
Look for	A mass of scarlet eggs on a leaf or stone near trees it favors (in spring).

BEST TOUCHY-FEELY LEGS
Daddy Long-Legs

spider

daddy long-legs

When you're hungry, you follow your nose to the kitchen. But when the daddy long-legs wants a meal, it follows its touchy-feely legs. Its second and longest pair of legs touches surfaces to sense food. When it finds something to eat, it tilts its body forward to see over its long legs and to test the object with its mouthparts.

pedipalps (leg-like mouthparts)

eight legs

cephalothorax

abdomen

makes silk

head, thorax, abdomen are all joined together

eight legs

can't make silk

Super Long Legs

A daddy long-legs has a pea-sized body with eight legs that can be one to two inches long. If you had the same proportions as a daddy long-legs, your legs would be forty or fifty feet long. You could step across the width of a basketball court!

BUG-FINDING TIPS

Key features	✓ pea-shaped body with long legs ✓ the size of a quarter (with legs) ✓ also called a "harvestman" ✓ predator or scavenger
Where	On grass, leaves, or rocks, in soil, or on your floor or wall.
When	Most active at twilight. Spring (eggs hatch), summer (nymph to adult), fall (eggs laid, adults die).
Look for	Teetering walk; fast moving.

MYTH OR FACT?
Daddy long-legs have a poisonous bite.
MYTH!
This bug helps people. Some kinds eat insects and others work on the clean-up crew, eating up decaying plants and animals.

SUPER SILK
Cat Face Spider

A female cat face spider weaves a big, beautiful orb web. Her life is like the famous Charlotte—weaving, trapping flying insects, mending, and finally tending an egg sac until she dies. She has so many superpowers that it's hard to pick just one, but let's look at silk threads.

Spider silk is a protein that's formed in a gland. The protein flows out a spigot (like a faucet) on the spider's spinneret (a silk-spinning organ). There, the protein changes shape from liquid to silk thread that she trails from her abdomen.

"First I dive at him. Next I wrap him up. Now I knock him out, so he'll be more comfortable . . . flies, bugs, grasshoppers, choice beetles, moths, butterflies, tasty cockroaches, gnats, midges, daddy long legs, centipedes, mosquitoes, crickets—anything that is careless enough to get caught in my web."

—*E. B. White, Charlotte's Web*

Seven Deadly Threads

Each spinneret has seven spigots for making different threads. Five are for weaving webs, one is for wrapping prey, and one is for bundling up egg sacs.

The strongest threads are called "dragline" silk. The spider uses these to weave the spokes and to anchor the web. She adds sticky "capture" silk to snare her victims. Both the dragline and capture silks are very stretchy, so the web doesn't break easily.

Why do spiders spin webs?
(Answer: Because they can't knit!)

Spider Jobs
Female spiders are larger than males so they can weave webs and lay eggs. The male has one function in life—to mate.

BUG-FINDING TIPS	
Key features	✓ marble-sized abdomen that resembles a cat's face ✓ has two bumps that look like a cat's ears ✓ predator
Where	Shrubs, sides of buildings, or porch lights.
When	Late summer to early fall (most noticeable when adult females have grown large).
Look for	Big, round web.

BEST AMBUSH
Antlion

A black and red ant scurries headlong into a cone-shaped pit, where a waiting antlion lurks, hidden in the sand at the bottom. As the ant tries to scramble back up, the antlion uses its shovel-like head to toss sand at the struggling prey. The ant loses its hold and slides down into the terrible jaws. The antlion then injects a dissolving juice, so it can suck up its meal.

TRY THIS: With a leaf stem, gently stir up the sand at the edge of a pit trap. Watch what happens.

"Pooh's first idea was that they should dig a Very Deep Pit, and then the Heffalump would come along and fall into the Pit."
—A. A. Milne, Winnie-the-Pooh

When you run up sand dunes, it's hard not to slip or tumble backwards— luckily there's no big-jawed beast waiting for you at the bottom!

Making a Pit Trap

The antlion digs backwards in a spiral, gaining traction with its bristly abdomen. The pit must be in a dry spot so that the sand or dirt will stay loose and keep a doomed insect from climbing back out. The antlion that digs a pit is just a youngster. That larva will become a winged insect that looks like a drab damselfly.

BUG-FINDING TIPS

Key features	✓ also called a "doodlebug" ✓ stubby legs ✓ big jaws ✓ half an inch in size ✓ predator
Where	Fine, dry soil or sand protected from rain; cave entrances, tree roots, or under porches.
When	Summer.
Look for	Cone-shaped pit traps about an inch across, often several together.

DEADLY HUG
Robber Fly

A robber fly perches in the tall grass, where it can spot bugs with its big eyes, then pursue its prey at high speed and strike in midair. This falcon of the bug world snags its victim with long, bristly legs— a deadly hug. Zooming back to its perch with its meal, the robber fly stabs the bug with its beaklike mouth and sucks up the bug juice.

Even tough beetles aren't safe from attack. Usually, a beetle's thick, scaly wings protect its body like a knight's armor. But when it flies, the beetle's soft abdomen becomes an easy target for a fly whose role as a robber is to steal the life of other bugs.

The Better to See You with, My Dear...

Those black, shiny eyes help the robber fly keep track of its fleeing prey. Bristles that look like a mustache protect the bulging eyes from struggling insects. Then comes the trapping embrace.

Waiter! Waiter! What is this bug doing in my soup?

(Answer: Hmm, I think that's the backstroke!)

BUG-FINDING TIPS

Key features	✓ bearded face ✓ big eyes ✓ females have a swordlike ovipositor for depositing eggs in cracks or in dead flower heads ✓ almost one inch long ✓ predator
Where	Open habitats in forests, grasslands, or along riversides.
When	Spring through fall.
Look for	Perched robber flies on shrubs, branch tips, or tall grass in sunshine, often near flowers that attract prey.

SUCKING PREY
Giant Water Bug

Walking along the water's edge of a slow-moving stream in western Montana, a fisheries biologist notices that the bank is lined with finger-sized, deflated balloons. He stoops down to pick one up. It's no balloon. It's a minnow with its insides sucked out. Nearby in the shallows, he finds the culprit: a giant water bug.

Rowing fast with its two pairs of hind legs, the bug grips the minnow with its front legs and stabs its beak deep to inject digestive juices. It hangs on until all the insides of the tiny fish have dissolved, then sucks in the liquefied organs. Minnows, frogs, tadpoles, and even small snakes have fallen victim to Montana's largest aquatic bug.

"He was shrinking before me like a deflating football. I watched the taut, glistening skin on his shoulders ruck, and rumple, and fall. Soon, part of his skin, formless as a pricked balloon, lay in floating folds like bright scum on top of the water. . . . an oval shadow hung in the water behind the drained frog; then the shadow glided away."

—*Annie Dillard,* Pilgrim at Tinker Creek

Digestive Juice Power

Bugs and people both put digestive juices to work. You just aren't able to inject them into another creature like a giant water bug does. Every time you eat, your digestive juices in your saliva and stomach break down your food into nutrients that your body can use.

BUG-FINDING TIPS

Key features	✓ nickname: "toe biter" (will bite if picked up) ✓ brown ✓ has front legs like pincers ✓ domino-sized ✓ predator
Where	Slow-moving streams, shallow ponds, and among underwater plants (hides in the mud). Flying adults are attracted to lights.
When	Summer, fall.
Look for	Deflated minnows or frogs or other evidence of grisly meals.

BIG-EYED BITER
Horse Fly

"But grandmother, what big eyes you have."
—Little Red Riding Hood

BZZZZZZZ. You're hiking up a trail on a sizzling July day with a horse fly buzzing around your head. If you slow down, she might be able to land and take a chunk out of you—ouch! It's hard to appreciate a biting fly, but notice those bright, big, shiny eyes—the better to see you with.

Only the female bites—that's so she can get the protein she needs to produce eggs. She chooses big mammals like horses, moose, and you. Her mouthparts tear skin and lap up blood. The male mouth is made to sip on flower pollen and nectar—helping to pollinate plants.

Eating with Scissors and a Straw

A female horse fly mouth is filled with blades that work like scissors to cut through your skin or the hide of an animal, such as a deer. Once the blood starts to flow, the horse fly sticks out her sucking mouthpart to slurp up her meal. She has nine different mouthparts that all work together. Although we say a horse fly "bites," she actually cuts and sucks.

BUG-FINDING TIPS

Key features	✓ brown ✓ big eyes ✓ twice the size of a housefly ✓ female is a predator ✓ male is a plant eater
Where	Forests, prairie, and near water and livestock.
When	Summer in full daylight without wind. August is a peak time.
Look and Listen for	Large fly buzzing around you, a horse, or a cow.

STINGING DEFENDER
Yellow Jacket

"But of course Turtle was slow, and one of the Yellow Jacket sisters stung him on the tail. Akee! Akee! Akee!"

—Fire Race: A Karuk Coyote Tale About How Fire Came to the People, *retold by Jonathan London*

If you swat or step on a yellow jacket (a kind of wasp), this fierce defender might sting you. Unlike a honeybee —whose stinger has a barb on the end, so it breaks off—a yellow jacket's needle-like stinger injects venom and comes out smoothly, so it can be used again and again. The venom causes swelling, pain, and itchiness.

Uninvited Guests

In early fall, yellow jackets love a good picnic that's loaded with soda, fruit, and cold chicken. Earlier in the summer, they hunt insects; however, when there are fewer insects to eat, the hungry yellow jackets head for our picnic tables for their protein and sugar. They're not out to harm you, but they can be annoying.

yellow jacket stinger

honey bee stinger

Stinger Defense

To defend a paper nest colony against a big animal such as a bear, it takes a powerful weapon. A yellow jacket uses its stinger—a hollow tube that connects to a venom sac inside its body—to inflict a painful wound. The yellow jacket curves its abdomen down, and the sharp stinger punctures the skin. Then it uses its muscles to drive the stinger in deep, while pumping venom from the sac into the victim.

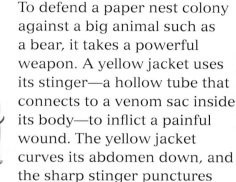

BUG-FINDING TIPS	
Key features	✓ yellow and black stripes ✓ not fuzzy like a honeybee but the same size ✓ omnivore
Where	Almost everywhere.
When	Spring through fall. Most active on warm days without rain.
Look for	Paper nests under steps, on shrubs, or under house eaves.

FAKE SNAKE
Tiger Swallowtail Butterfly Caterpillar

A fat caterpillar is a favorite bird snack. To keep predators away, the tiger swallowtail caterpillar has two yellow eyespots on its head to make it look like a scary snake. If a ravenous bird gets close, the fake snake may rear up in a threatening pose. When attacked, the swallowtail caterpillar shoots out a startling orange fork from its head that releases a stinky smell.

What does a caterpillar do on New Year's Day?

(Answer: Turn over a new leaf!)

One of Montana's largest butterflies: The western tiger swallowtail flits across many parts of western and central Montana—in both Glacier and Yellowstone national parks. Look for them flying on hilltops and ridges to search for mates.

Munching Machine

A caterpillar eats twenty times its weight in plants by the time it's ready to spin a chrysalis. If you ate like a caterpillar, you'd have to chow down fifty pounds of food a day. You wouldn't have time to hide or run from predators—better to just have a pair of scary eyespots so you can keep on eating.

BUG-FINDING TIPS	
Key features	✓ green to blend with leaves ✓ yellow eyespots ✓ up to two inches long ✓ plant eater
Where	Woods near rivers and streams, canyons, parks, roadsides, and trees in backyards.
When	Summer.
Look for	Caterpillars on ends of leaves of cottonwood, aspen, chokecherry, ash, and willow trees.

BAD ODOR POWER
Green Stink Bug

If you pick up a stink bug and hold it between your thumb and forefinger, it will ooze a smelly liquid. Yuck! That odor cuts down on attacks from birds, toads, big spiders, and kids, too. The green stink bug's defense comes from glands on both sides of its thorax, located between the second and third pair of legs. A gland is an organ that produces a special substance and releases it. You have glands in your mouth that make saliva.

Secret Weapon Facts

● This insect has another name: shield bug (for its shape). Just think if medieval knights had shields that sent out a stinky smell when struck. That would have made for some skunky skirmishes.

● A stink bug's chemical brew takes up 5 percent of its body weight. That would be like you carrying a one-gallon bucket of goo around.

● Its chemicals are called "aldehydes" (AL-duh-hides). In low concentrations, they might be almost pleasant smelling, but the stink bug makes a powerful, nasty concoction.

● Smelly fumes may also serve as an alarm signal to other stink bugs, as if to say, "I'm under attack. Scatter!"

What did the judge say when the stink bug walked in?

(Answer: Odor! Odor in the court!)

BUG-FINDING TIPS	
Key features	✓ shield-shaped, bright green with yellowish or reddish edges ✓ the size of a small paperclip ✓ plant eater
Where	Gardens, fields, and orchards.
When	Spring (nymphs), summer (adults).
Look for	Flying around lights; feeding on flowers, leaves, or fruit; green egg clusters on undersides of leaves.

EYESPOT TRICKERY
Common Wood Nymph Butterfly

A drab butterfly rests on a branch as the morning sun warms his wings so he can fly. The dark brown color helps him blend in with the tree shadows. The two eyespots on the underside of his forewings are tucked out of view. Just then, a hungry yellow warbler spots the wood nymph. The butterfly flashes his wings to reveal alarmingly big and realistic eyes. The songbird veers off, fooled by the fakery.

How Eyespots Work

Does a songbird veer away because a butterfly's eyespots look like the eyes of a hawk? The latest research suggests that may not be the case. Eyespots work no matter what shape they are as long as they are startling—like wearing a mask and jumping out from behind a tree and yelling, "boo!"

BUG-FINDING TIPS

Key features	✓ brown ✓ two big eyespots on the forewings ✓ lower hind wing has smaller eyespots ✓ wingspan of up to two inches ✓ eats nectar (adult) ✓ eats grass (caterpillar)
Where	Sunny meadows, forest openings, prairies, and lawns.
When	Summer: July, August (peak). Caterpillars hatch in fall, hibernate, then feed the next spring.
Look for	Slow, dipping flight throug grasses and flowers.

WASP MIMIC
Hover Fly

If you're going to be a master of disguise, you might study the hover fly. This fly looks and acts like a wasp. A wasp has long antennae, but a hover fly has short ones. So this fly waves its two front legs in front of its head to mimic long antennae.

Why would a fly want to disguise itself as a wasp? The hover fly is harmless and spends its time feeding on and pollinating flowers, giving it another name, "flower fly." Without a stinger to defend itself, the fly might be more likely to be eaten by birds or crab spiders. Copying a dangerous wasp can fool predators into leaving it alone.

Flies in Disguise

Flies have two wings. Yellow jackets, a type of wasp, have four wings. To fake a yellow jacket's wings takes another tricky disguise. A real yellow jacket at rest folds its wings lengthwise along its body. The hover fly can't fold its two wings, but the brown bands along the leading edges of its transparent wings look like folded wings.

HOVER FEAT
Hover flies seem to hang in the air as if suspended by magic, instead of by buzzing wings. Some male hover flies can stay in one place in the air for hours on sunny days, probably waiting to attract a female mate.

What do you call a fly without wings?
(Answer: A walk!)

yellow jacket

hover fly

BUG-FINDING TIPS

Key features	✓ the same size and appearance as a wasp ✓ eats nectar, pollen, and aphid honeydew ✓ some look like fuzzy bumblebees
Where	Openings in forest areas.
When	Spring, summer.
Look for	Hovering in flight.

MOSQUITO LOOK-ALIKE
Giant Crane Fly

Stop! Don't swat that long-legged, huge monster mosquito on the wall. That mouth isn't for sucking blood. Those skinny, wobbly legs aren't for landing on you. They're for grabbing hold of long grass.

mosquito   crane fly

Mosquito versus Crane Fly

Take a close look at the two side by side. What's the same? What's different? Both are flies, so they have two wings, but they are in different families. Both have long legs, but the mosquito's legs are shorter. A mosquito is a good flier, and a crane fly is a shaky one. A female mosquito has a mouthpart for biting and sucking blood, but the giant crane fly mouth isn't for eating at all! Only the wormlike larvae eat—scavenging on decaying plants.

GOLLYWHOPPER

Crane flies live all over the world. They are so eye-catching that people have come up with nicknames that often are based on myths, such as "skeeter eaters," "mosquito hawks," "gully nippers," "daddy long-legs," "jimmy spinners," and "gollywhoppers."

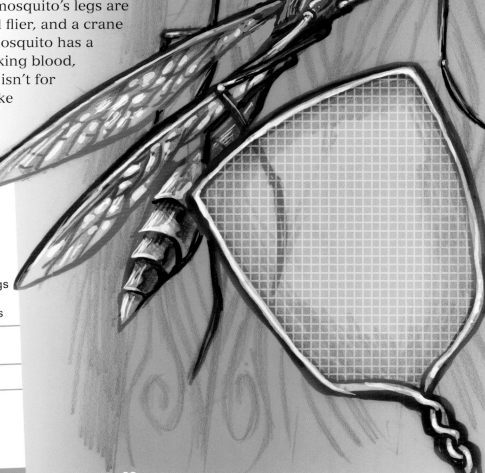

BUG-FINDING TIPS

Key features	✓ largest fly west of Rocky Mountains ✓ wingspan of up to three inches ✓ white racing stripes on its thorax ✓ clear wings and long legs ✓ adults don't eat and only live about five days
Where	Humid places, edges of forests, and near streams.
When	Spring, summer.
Look for	Activity around lights at night—outdoors and indoors. During the day, they're hard to find in grass and leaves.

HUMMINGBIRD COPYCAT
White-Lined Sphinx Moth

Is it a bird? Is it a helicopter? Is it a ...sphinx moth? Also called a "hawk moth" or "hummingbird moth," this big, furry moth acts like a hummingbird, hovering while sipping flower nectar. Instead of a tongue, it uncurls its long, hollow mouthpart called a "proboscis" (pruh-BOSS-kiss) to reach deep into a flower.

The sphinx moth flaps its narrow wings in strong, swift beats. Watch one pause in midair and swing side to side quickly—called swing-hovering. This might serve as a ploy to dodge birds or bats looking for a mouthful.

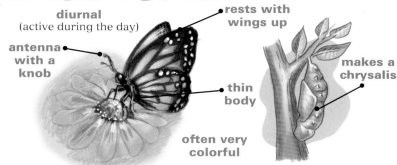

diurnal (active during the day)

antenna with a knob

rests with wings up

thin body

often very colorful

makes a chrysalis

butterfly

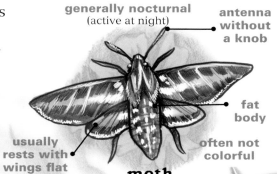

generally nocturnal (active at night)

antenna without a knob

fat body

usually rests with wings flat

often not colorful

makes a cocoon

moth

STRANGE BUT TRUE
The white-lined sphinx moth sips nectar, but four of the six kinds of sphinx moth adults in Montana don't eat at all. Their only purpose? To mate and lay eggs.

White-lined Sphinx Moth Caterpillar
The big green caterpillar with a short "horn" on its back end can pose like the ancient sphinx—holding its legs up and tucking its head.

BUG-FINDING TIPS

Key features	✓ pale tan stripe from the base to the tip of its wing ✓ thick body ✓ two-inch wingspan ✓ cocoons are underground ✓ eats nectar
Where	Many habitats, from mountains to prairies.
When	Summer; flies day and night. Evening is the best time to see them around.
Look for	Evening primroses and flowers in the sage family.

HONEYDEW MAKER
Aphid

In the insect world, aphids are the cows and ants are the dairy farmers. Instead of serving up creamy milk, aphids ooze honeydew. On the underside of a leaf, ants cluster around tiny aphids. With a hand lens, you can see the ants milking the aphids for a sugary drink. The ants, in turn, drive off predators like ladybugs and lacewings. Not all aphids are so lucky to have ant protectors. One study found that 63 percent of aphids without ant guards were eaten by predators; only 26 percent of aphids with ant guards were eaten.

Sap Sipper

An aphid has a sweet tooth. With its piercing mouthpart, the bug sucks sap from plant stems and leaves. The honeydew is actually the poop that an aphid excretes out the end of its abdomen. The tasty droplets attract hungry butterflies, flies, bees, wasps, and ants.

MULTIPLYING APHIDS

A female aphid can have a baby that's already carrying another baby inside of her. That means a newborn aphid will have her own baby in a few days. The mom doesn't even need to mate with a male first. However, by summer's end, some females do mate and lay eggs that will hatch the following spring.

BUG-FINDING TIPS	
Key features	✓ oval-shaped and often green ✓ tiny, the size of a pinhead ✓ plant eater
Where	Gardens, leafy plants.
When	Early summer is best, when there are plenty of new leaves.
Look for	Wilted or curled leaves at the end of a branch or stem. Ants on a plant stem are a good sign.

NECTAR COLLECTOR
Bumblebee

"To make a prairie it takes a clover and one bee, One clover, and a bee, And revery . . ."
—Emily Dickinson

A bumblebee gathers sweet nectar and pollen from a wild rose to take back to the larval bees. Before heading back to the nest, it bumbles off to another rose. Pollen from the first rose's anther (male part) falls onto the stigma (female part) of the second rose—voilà, the female is fertilized. Many wildflowers need pollinators like bees, birds, butterflies, moths, or bats to fertilize them so they can reproduce. So please, *bee* kind to bees!

Dumbledore Bumbledore!

Dumbledore means bumblebee in Old English. J. K. Rowling, author of the *Harry Potter* books, chose the name because she pictured the Hogwarts' headmaster wandering the halls humming tunes to himself like a bee.

What kind of bee has no stinger and no wings?
(Answer: A Fris-bee!)

BUG-FINDING TIPS

Key features	✓ fuzzy ✓ black and yellow/orange ✓ will sting ✓ not aggressive ✓ marble-sized ✓ eats nectar and pollen
Where	Flowers, nest sites (vacant rodent burrows or small underground tunnels).
When	Spring through fall.
Look for	Impressive humming buzz; note pollen loads in leg baskets.

wings
Vibrating flight muscles make the buzz, plus help a bumblebee warm up on a chilly day.

long tongue
For reaching the flower's nectar.

Busy Bee Body

A bumblebee is equipped for collecting nectar and pollen, and for pollinating, too.

pollen basket
For storing pollen.

fuzzy body
When a bumblebee flies, the hairs create static electricity (like rubbing your hair with a balloon); this charge helps the pollen grains stick to the body and pollinate more flowers.

honey stomach
A bag inside its abdomen for storing nectar.

BIGGEST APPETITE
Ladybug

A ladybug grips the underside of a leaf and marches on sticky pads toward a cluster of aphids. With no ants in view, the beetle barges in for a meal like a kid in charge of the candy store. One ladybug may devour sixty aphids a day. The larvae eat aphids, too.

Gardeners feel lucky when these aphid-munchers arrive. Sometimes, people buy ladybugs for their gardens, but most of them will fly off. It's better to let the native ladybugs find the aphids—and they will.

Ladybug, Ladybug Fly Away Home

You've just hiked to the top of a peak in Glacier National Park in August. Your family sits down on boulders to eat lunch. Then, you see it: a mass of shiny red beetles packed tight in a crack. Why would ladybugs gather high on a peak? Mountaintops are places where it's easy for ladybugs to find each other so they can mate before winter. Grizzly bears will climb up to peaks, too, for a ladybug meal. They don't seem to mind the bad-tasting toxic fluid these bugs secrete from the joints in their legs.

BUG-FINDING TIPS

Key features	✓ count the spots: Montana natives include the two-spot and the thirteen-spot ladybug; a species brought here from Europe has seven spots ✓ pea-sized ✓ predator
Where	Leafy plants, bushes, and mountaintops.
When	Summer through fall. In winter may snuggle in house nooks—waking up when too wam.
Look for	Aphids and ladybugs on undersides of leaves.

GREEDY GOBBLER
Green Lacewing

Here's a bug with an insatiable appetite. The larva of the green lacewing skewers aphids with jaws that are like a pair of tongs. After piercing and sucking its victim dry, the predator moves on to the next aphid or other soft-bodied insect. Some kinds of young lacewings will disguise themselves by piling aphid carcasses on their bristly backs. That way, they can sneak by ants that often guard aphids. While they look more like alligators, the larvae are nicknamed "aphid lions" for their ferocious behavior.

Gardener's Friend

Lovely, adult green lacewings flitting in a garden offer more than a pretty sight. Gardeners adore lacewings because both adults and larvae devour aphids that eat their crops. Planting pollen and nectar flowers in a vegetable garden encourages lacewings to live there.

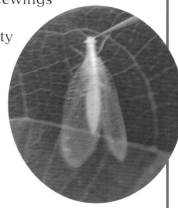

adult green lacewing

Cannibals!

When the first lacewing larva hatches, if it had a chance it would eat up its brothers and sisters before they could wriggle free of their eggs. To prevent cannibalism, the female separates the eggs from each other. She lays an egg on the end of a thin silk stalk she makes and attaches to the underside of a leaf. Then she lays the rest of her eggs in the same way. These isolated eggs are also harder for predators to find.

lacewing larva and aphids

BUG-FINDING TIPS

Key features	Larva: ✓ brown, alligator-shaped ✓ aphid predator Adult: ✓ green, lacy, clear wings and golden eyes ✓ the size of a small paperclip
Where	Gardens, leafy plants and shrubs, and porch lights.
When	Spring, summer.
Look for	Adults landing on you in your garden. To find larvae, look for aphids.

BARK BATTLER
Mountain Pine Beetle

A lodgepole pine battles for its life. Will the mountain pine beetles bore through the bark to reach the fleshy layer underneath? Will the tree be able to trap the beetles in sticky pitch and push them out?

At first, the tree looks like the winner, pitching out one beetle after another. But these bugs release chemicals called "pheromones" to signal for more beetles to join the front lines. Sure enough, reinforcements come flying in. Males and females bore holes from all sides. The 100-year-old tree cannot defend itself. Once inside, the beetles employ their next tactic. They've brought a blue fungus that will clog the tree's water supply—and they've won.

Warming Climate— More Beetles

Cold is the beetle's worst enemy. As our climate warms, mountain pine beetles live longer and survive higher up on mountains, invading higher-elevation trees like whitebark pines.

TUNNELING CLUES

Adult beetles chew long tunnels into the tree, and females make tiny side chambers where they lay their eggs. When the larvae hatch, they eat both wood and the nutritious blue fungus as they bore their own tunnels. They turn into pupae and later emerge as adults, who fly away.

BUG-FINDING TIPS

Key features	✓ black ✓ the size of a match head ✓ plant eater
Where	Pine forests (especially lodgepole) with trees more than five inches in diameter
When	Adults emerge in midsummer
Look for	Reddish needles on infected trees, globs of pitch on the bark, sawdust at the tree base, tunnels in tree layer under the bark, blue fungus on stumps and woodpecker holes.

WOOD EXCAVATOR
Carpenter Ant

Unlike human carpenters, these ants build *in* wood, not *with* wood. They bite through the soft wood of decaying or dead trees to make tunnels for nests. Unlike mountain pine beetles, carpenter ants don't actually eat the wood. Instead, they munch on living or dead bugs.

If you could shrink to ant size, you'd creep through a maze of tunnels that lead from your parents' nest to smaller outlying nests. Along the way, you'd meet thousands of family members. A huge colony may have 100,000 ants with several queens.

Tunnel Living

Wingless female workers take care of the tunnels and serve as nannies to the eggs, larvae, and pupae (which are like cocoons, where larvae transform into adults). Winged fertile males and the females fly off to mate in midair and start new colonies.

After mating, the males die and the females lose their wings before excavating new homes. It takes three to six years for ants to form a large colony.

Beware the Pileated Woodpecker

Watch out! Here comes that long, sticky, barbed tongue. Nailed! There goes another carpenter ant. Montana's biggest woodpecker feasts on the state's biggest ant.

BUG-FINDING TIPS

Key features	✓ black ✓ slow moving ✓ no sting ✓ stong jaws for biting wood ✓ the size of a small paperclip ✓ predator
Where	Forests with dead or dying trees; can be in houses with rotten wood.
When	Early summer for winged adults. Year-round in trees, but not active in winter.
Look for	Sawdust piles at the base of dead or downed trees, big woodpecker holes.

BOOKS

Acorn, John. *Bugs of Alberta.* Lone Pine Publishing.
Lively descriptions of 125 bugs in Alberta, Canada, just to the north of Montana—so it works well for Montana, too.

Kaufman, Kenn and Eric R. Eaton. *Field Guide to Insects of North America.*
Houghton Mifflin Company. A terrific field guide with interesting natural history notes. Easy to use.

Stokes, Donald. *A Guide to Observing Insect Lives.* Little, Brown and Company.
A hands-on guide to investigating bugs in every season, starting in your own backyard.

Golden Guide books from St Martin's Press.
Look for these pocket-sized, colorful books on *Insects, Butterflies and Moths,* and *Spiders and their Kin.*

WEBSITES

www.enature.com
Sights and sounds of 6,000 species.

http://ufbir.ifas.ufl.edu/
University of Florida Bug Records site. Find out the fastest, largest, loudest, and more.

http://teacher.scholastic.com/activities/bugs/
Play Monster Bugs. Build a caterpillar. Look up cool facts about bugs.

http://www.sciencenewsforkids.org/
Keep up on the latest science discoveries, including bugs. Type "bug" into the search box.

www.earthsky.org/kids
The popular "Earth and Sky" radio program has a kids' website that will lead you
to some great bug programs. Check out the questions posed by kids to scientists.
Type "bug" into the search box.

http://www.royalalbertamuseum.ca/natural/insects/bugsfaq/most.htm
Fabulous bug facts and photos.

ABOUT THE AUTHOR AND ILLUSTRATOR

Deborah Richie Oberbillig writes about nature from her Missoula, Montana, home that she shares with her husband Dave, son Ian, dog Luna, gopher snake Slither, and gecko Blue Fire. She credits her love of the natural world especially to her father, Dave Richie. She has a master's degree in journalism from the University of Montana–Missoula and a bachelor's degree in biology from the University of Oregon. She is the author of *Bird Feats of Montana.*

Robert Rath is a book illustrator and designer living in Bozeman. Although he has worked with Scholastic Books, Lucasfilm, and Montana State University, his favorite project is keeping up with his family.